Process Development Symposium Program

Held at the 2023 AIChE Spring Meeting and 19th Global Congress on Process Safety

Houston, Texas, USA
12-16 March 2023

ISBN: 978-1-7138-7940-4

Printed from e-media with permission by:

Curran Associates, Inc.
57 Morehouse Lane
Red Hook, NY 12571

Some format issues inherent in the e-media version may also appear in this print version.

Printed with permission by Curran Associates, Inc. (2023)

For permission requests, please contact AIChE
at the address below.

AIChE
120 Wall Street, FL 23
New York, NY 10005-4020

Phone: (800) 242-4363
Fax: (203) 775-5177

www.aiche.org

Additional copies of this publication are available from:

Curran Associates, Inc.
57 Morehouse Lane
Red Hook, NY 12571 USA
Phone: 845-758-0400
Fax: 845-758-2633
Email: curran@proceedings.com
Web: www.proceedings.com

TABLE OF CONTENTS

PROCESS INTENSIFICATION IN INDUSTRY

Sustainability in Modular Process Plants

David Edwards

Sustainability in the CPI typically focuses on making the process chemistry more benign, or using renewable sources of raw materials, or how products can be designed with recycling in mind. But what about the plant itself? What is being done to make the design, fabrication, installation, operation and eventual decommissioning of process plants more sustainable? This article will focus on modular process plants, and what work is underway to reduce the environmental footprint in their design and fabrication, offering a comparison to the traditional stick-built approach.

A Unified Sustainability Analysis Framework to Advance Circular economy for Post-consumer Plastic

Rui Shi

Plastic disposal has been generating serious environmental problems in the past decades. Currently, only 9% of plastic waste is mechanically recycled in the United States. Chemical recycling breaks down the *plastics* waste into its constituents, therefore complements mechanical recycling, reducing the environmental impacts and resource depletion. Processes such as pyrolysis, hydrothermal liquefaction, etc., can handle hard-to-recycle plastics, while producing a wide range of valuable products including oil, waxes, and aromatics, providing opportunities to reduce oil usage, carbon dioxide emissions and the quantities of waste for disposal. On the other hand, the economic potential and environmental impacts have not well studied for particular feedstocks and pathways.

Here, we focus on design, synthesis, and simulation of chemical recycling process of waste feedstock, and examine the sustainability implications against waste to energy and mechanical recycling. For this study, we investigated multiple advanced recycling processes, conducted life cycle and techno-economic analysis to characterize the economic potential and environmental impacts, and elucidated the challenges in commercializing the technology due to the impact of operating capacity, capital cost, and feedstock, etc. Ultimately, this work demonstrates a clear need to leverage integrated life cycle and techno-economic analysis modeling platforms to advance plastics recycling systems as a whole, and specific recommendations are made for the prioritization of development pathways to achieve circular economy of plastics.

Sustainable Energy Transition

Sherard Sadeek, Keeran Ward, Dhurjati Prasad Chakrabarti

With a heavy reliance on the oil and gas industries, Trinidad and Tobago is one of the countries with the highest per-capita CO2 emissions in the world. Rapid redesign of Trinidad and Tobago's power sector is essential to fostering a sustainable energy transition given the nation's commitment, as a small island developing state (SIDS), to sustainable development goals and climate change agreements. The current study uses a mixed-integer linear programming model (MILP) to analyze the minimization of energy costs (LCOE) and greenhouse gas emissions (GHGLC) across 5 scenarios characterizing the national power generating system. Our calculations indicate that current take-or-pay power purchasing agreements (PPAs) aligned to 'Business-as-Usual' (Single cycle and Combined cycle) have a take-or-pay LCOE of $USD 61.72 per MWh and a GHGLC of 756 kgCO2eq per MWh. Combined advancements in resource efficiency and power generating technology (Combined cycle and Combined cycle) led to reductions in LCOE and GHGLC of up to 40% ($36.94 USD per MWh) and 24% (572 kgCO2eq per MWh), respectively. Furthermore, despite possible saved emissions when compared to local business-as-usual operations, the introduction of low carbon H2, MeOH, and NH3 into the process system was not financially feasible at this time. Finally, looking ahead to 2030 reveals an overall increased LCOE of ($USD 8.5–14) per MWh and GHGLC of 8–27 kgCO2eq, both of which are related to increased grid demand and natural gas pricing. Research findings thus indicate the requirement for active decarbonization of the domestic energy sector and the repeal of existing regulations that hinder the use of low-carbon technologies for sustainable development

Direct Injection Liquid Chromatography (DILC) Enables Rapid Process Intensification

Jason Hein

Reaction discovery, optimization and understand critically relies of making measurements of the total composition of a reactive system. In-situ reaction kinetic measurements represent the gold standard for deconvolution of a reaction network. This requires the extraction of real-time reaction profiles detailing the time-resolved concentrations of reagents, catalysts, ligands, intermediates, products and by-products using on-line or in-line process analytical technologies such as HPLC, NMR and FTIR. However, many common reactions contain physical and chemical challenges which render the direct, data-dense extraction of this information untenable. In particular, systems that are heterogenous, air-sensitive, or dangerously reactive are particularly intractable, rendering them opaque to mechanistic analysis using process analytical technology.

Our group specializes in the application of automation and custom analytical hardware to visualize critical mechanistic detail from some of the most challenging reaction environments. This presentation will introduce some of the newest technology created in our group, including tools to enable online visualization of solid-liquid and liquid liquid heterogenous reactions. By applying these tools, mechanistic details can be rapidly elucidated, facilitating reaction optimization and process validation in dramatically reduced timelines.

Development of a Novel Aromatic Extraction Process Design for Low Aromatic Content Feeds

Adam McNeeley, Chang-Che Tsai, Y. Liu, Shiang-Tai Lin

Ionic liquids (ILs) as chemical solvents offer unique properties such as non-volatility and tunability that are attractive have potential to displace conventional organic solvents. The prevalence of aromatic moieties in IL cations make them favorable for the selective extraction of aromatic species from hydrocarbon mixtures. A particular aromatic extraction application of interest is the naphtha feed of low aromatic contents for ethylene cracking processes, which cannot be economically separated using commercial technologies and have been shown to provide cost and energy advantage in the downstream ethylene cracking process.

Prior investigation on this topic has demonstrated how to effectively model thermodynamic behavior of IL-hydrocarbon systems, and the design and optimization of an IL-based aromatic extraction process. Given the extremely large number of feasible IL solvents, our goal is to understand how the properties of an IL affect the performance of the process. We identify 16 well studied ILs with available experimental data and produce optimized designs for each IL and a comparative example using sulfolane solvent.

We compare process results to conventional IL solvent attributes such as distribution and selectivity and introduce a new science-guided solvent attribute. We perform partial least squares (PLS) analysis to determine the solvent attributes most correlated with process steam duty. The low aromatic feed composition cases show process steam duty to be highly correlated with the new solvent heat load attribute introduced in this work. We also find key process variables expected to influence equipment size of major units to be correlated with the solvent heat load variable. 11 of the 16 studies ILs show reduction in process steam duty compared to sulfolane with the best performing IL required only 57% of process steam duty compared to sulfolane.

Thermochemical Production of Green Hydrogen or Syngas

Kent J. Warren, Britt Boughey

Meeting growing domestic energy demands while reducing our nation's carbon footprint requires the development of technologies that not only harness the abundant renewable resources of the United States (e.g., concentrated solar radiation and renewable electricity) but also allow for such energy to be dispatched on demand to the end user, regardless of the time of day or geographic location. One particularly attractive approach is the production of syngas via thermochemical water and carbon dioxide splitting, as such pathway – unlike alternative electrolysis-based technologies – relies entirely on the conversion of process heat which better complements traditional economies of scale. Importantly, syngas, when integrated with well-established catalytic processes like Fischer-Tropsch synthesis, can be converted into stable, transportable hydrocarbons that are compatible with existing infrastructure.

In this talk, we will present an innovative thermochemical process using a novel metal oxide that enables breakthrough energy efficiencies in the production of syngas or green hydrogen. This process operates isothermally, eliminating more than 60% of the losses faced by current state of the art thermochemical water or carbon dioxide splitting technologies that incorporate wide temperature swings. If operating solely from electrical heat, even higher efficiencies can be realized as re-radiative losses associated with concentrated solar thermal heat sources would be eliminated. The opportunity to provide process heat via low-cost and low-carbon sources of electricity also unlocks the potential for a reactor system that can be co-sited with other industrial processes, potentially scavenging waste heat to comprise some of the process heat requirements and thus further lowering OPEX.

Opportunities for Process Development Engineers in Accelerating the Transport Sector Energy Transition

Shyamal Bej

Opportunities for Process Development Engineers in Accelerating the Transport Sector Energy Transition

Shyamal Bej

Abstract:

The energy mixture presently used to fuel the transport sector is predominantly of fossil origin and leads to significant net carbon dioxide emissions. As a result, there is an urgent need to transition the existing energy mixture into one which consists of renewable sources and has minimal fossil origin. After giving a brief overview of the history of previous energy transitions and the vast amount of knowledge generated while working through these transitions, this talk will focus on future fuels proposed for the transport sector and the technologies required to produce these future fuels. A summary of examples where knowledge gained during earlier transitions can be leveraged and where new technologies need to be developed will be reviewed. Enormous opportunities exist for process development engineers in improving current technologies as well as developing new technologies for successfully achieving the goals of the current energy transition.

Assemblytron: Flexible Automation of DNA Assembly with Opentrons OT-2 Lab Robots

John Bryant Jr.

As one of the newest fields of engineering, synthetic biology relies upon a trial-and-error Design–Build–Test–Learn (DBTL) approach to simultaneously learn how a function is encoded in biology and attempt to engineer it. Many software and hardware platforms have been developed to automate, optimize and algorithmically perform each step of the DBTL cycle. However, there are many fewer options for automating the build step. Build typically involves deoxyribonucleic acid (DNA) assembly, which remains manual, low throughput and unreliable in most cases and limits our ability to advance the science and engineering of biology. Here, we present AssemblyTron, an open-source Python package to integrate j5 DNA assembly design software outputs with build implementation in Opentrons liquid handling robotics with minimal human intervention. We demonstrate the versatility of AssemblyTron through several scarless, multipart DNA assemblies, beginning from fragment amplification. We show that AssemblyTron can perform polymerase chain reactions across a range of fragment lengths and annealing temperatures by using an optimal annealing temperature gradient calculation algorithm. We then demonstrate that AssemblyTron can perform Golden Gate and homology-dependent *in vivo* assemblies (IVAs) with comparable fidelity to manual assemblies by simultaneously building four four-fragment assemblies of chromoprotein reporter expression plasmids. Finally, we used AssemblyTron to perform site-directed mutagenesis reactions via homology-dependent IVA also achieving comparable fidelity to manual assemblies as assessed by sequencing. AssemblyTron can reduce the time, training, costs and wastes associated with synthetic biology, which, along with open-source and affordable automation, will further foster the accessibility of synthetic biology and accelerate biological research and engineering.

Engineering Low Cost Nanotextured Steel for Protein Therapeutic Delivery

Maeve Janecka

Nanoneedles are a minimally invasive platform that contains nanoscale spikes which enable access to the intracellular environment. Nanoneedles can be engineered to deliver biological molecules and nanoparticles across the cell membrane, and to probe cytoplasmic fluids for biomarkers of diseases such as cancer. The physicochemical interface of these nanoneedles is tunable and an ideal candidate for intracellular delivery. However, current fabrication methods require silicon lithography, which has limited scalability due to its high cost.

Stainless steel is commonly used for biomedical implants due to its mechanical strength, corrosion resistance and biocompatible properties. We have developed an ultra-low cost and industrially relevant process to modify the surface of SS316L stainless steel through chemical nano-etching, resulting in stainless steel nanotextures that resemble solid porous silicon nanoneedles. The nanotextured steel has been shown to be non-cytotoxic to fibroblast mammalian cells (NIH-3T3), anti-bacterial to both gram-positive and gram-negative bacteria, and maintains all the physical properties of stainless steel. We evaluated the loading capacity of recombinantly expressed sfGFP +10 and -10 charged variants on engineered nanotextured steel. HeLa cells plated on sfGFP-coated nanotextured steel suggest improvement in cellular uptake compared to unmodified SS316L stainless steel, assessed via flow cytometry. In the future, we will evaluate the efficacy of siRNA delivery to the cytosol to evaluate the feasibility of delivering functional proteins. Low-cost nanotextured steel can be further expanded to deliver proteins and genes to promote increased vascularization and perfusion to orthopedic implant sites, improving surgical recovery times.

Solving Equipment Specific Knowledge and Skills Gaps with Knowledge Automation

Christian McDermott, Francis Montemurro

US Manufacturing is at an inflection point: it is increasingly crucial to economic growth yet plants are struggling to attract and retain a skilled workforce. This is a top priority for almost all manufacturing executives.

The traditional way of onboarding new hires and relying on a few SMEs (Subject Matter Experts) on critical operations is not working.

This session will look at how Knowledge Automation, deployed by the world's biggest companies, is proven to solve this challenge, leading to safer, more productive operations thanks to procedure clarity and knowledge capture.

We will show real examples and data on how a Knowledge Automation platform will:

- Capture & transfer critical operator knowledge
- Provide procedure clarity in operations and maintenance
- Make onboarding efficient and effective
- Prevent operations personnel (SMEs) being overwhelmed
- Reduce unplanned events and incidents

Scale-Down and Scale-up of a Slurry Mixing Process

Richard Grenville, Jason Giacomelli, Benjamin Boyer

Scale-up is a common activity in process development. The goal is to develop the understanding of how a particular unit operation affects the outcome of the process in small scale equipment and then design equipment for the full-scale plant based on this understanding.

If a piece of equipment in a plant is not operating as designed, the operation can be scaled down, recreating the problem at small scale to develop an understanding of its causes. Then further tests can be carried out to develop a solution to the problem at small scale followed by the design of equipment to implement at plant scale.

A case study will be presented which will discuss how mixing issues were created, due to changes in the operating procedures, in the production of a viscous, non-Newtonian, high solids concentration slurry. The changes resulted in the slurry being more viscous and shear-thinning. The agitator was not capable of moving the slurry in regions far from the impeller blades and severe stagnation and fouling was observed. The process would have to be shut down on a regular basis so that the vessel could be cleaned.

Scale-down studies were carried out in a combined stirred tank / viscometer developed for the project enabling a mixing test to be performed identifying the degree of agitator required to eliminate stagnant regions and, at the same time, to measure the slurries rheology. The results were scaled-up and the new agitator has been working as specified for nearly five years.

Finally, some of the considerations that must be taken into account when making rheology measurements, especially for high concentration slurries, will be addressed.

Never Let a Good Crisis Go to Waste: RAPID RACER AMPD and the Opportunity to Accelerate Process Development – An Update

Patricia Gillenwater

As announced on March 7, 2022, the RAPID Manufacturing Institute® (an AIChE entity and one of the Manufacturing USA Institutes) was awarded $4.64MM by the U.S. Dept. of Commerce's National Institute of Standards and Technology (NIST) to both develop a methodology to accelerate process development and also develop a testbed for domestic manufacturing of critical pharmaceutical ingredients. The testbed will be located in the underserved Appalachian region in South Charleston, WV. The project is funded under the American Rescue Plan Act of 2021, also known as the COVID-19 Stimulus Package.

The test bed program, currently under development, presents an opportunity to accelerate chemical process development with modular skids and demonstrate its potential for rapid deployment of distributed manufacturing. Work is also underway to deliver an accelerated process development methodology to support the reshoring efforts that may take place during the next crisis/pandemic. We will present the learnings from our benchmarking survey of the AIChE Process Development Division on the current state of process development and how they are being codified into a stagewise methology for accelerated process development. Current work being done to explore the feasiblity of using a Damkohler number as a potential unifying tool for both the traditional reactor application as well as extension to downstream processing will also be discussed. An update on the test bed program will also be provided.

Batch-to-Continuous Conversion of Substituted Alkyl Phenols

Nasser Al Azri, Glenn Cormack, Nico Proust, Zack Palchak, Michael DeLetter

Gasoline additives are a critical element of the global transportation fuels industry, specifically in North America where light passenger vehicle fuel requires mandatory additization. The global market for gasoline additives is circa 190,000MT/yr and is predominantly composed of three different chemistries: substituted alkylphenols, polyisobutylene (PIB) amines, and polyether amines. The premise of this project is to support market growth by minimizing capital and operating expenses of new capacity through modular chemical process intensification and batch-to-continuous conversion for production of substituted PIB phenol additives, known as Mannich detergents.

The current manufacturing process for the production of Mannich detergents involves the reaction of a PIB phenol with formaldehyde and a low molecular weight amine. The current batch process is executed in a 30MT reactor and entails 7-8 complex steps resulting in a 40-hour cycle time. The main driving forces for the long cycle time are mainly due to heat and mass transfer limitations associated with the high viscosity of the PIB at reaction temperature, the presence of a volatile amine above its vapor pressure, the presence of water as a by-product and diluent, and the need for exothermic reaction control. Currently, to resolve the viscosity limitations, a solvent is added which creates additional complexity during the post-reaction distillation of water and other volatiles.

In this work, the authors will present a case study on practical lessons learned through lab and pilot plant scale up of an intensified methodology for production of Mannich detergents. The study includes the minimization of heat and mass transfer limitations through novel unit operations, and the development of a kilo-scale continuous manufacturing unit. Finally, the authors will discuss practical considerations for modular unit scale up and deployment including OSBL infrastructure, raw materials, unit siting, controls integration, intermodal shipment, and financial analysis.

Chemical Transformers and the Clean Hydrogen Grid

Robert Wegeng

Compact Hydrogen Generators, placed as "chemical transformers" on the natural gas grid, can be compared to electrical transformers on the electrical grid. Reconditioning the natural gas (or renewable natural gas) supply to produce hydrogen at or near points-of-use, is a near-term opportunity to begin converting the existing gas grid infrastructure into a highly functional, clean hydrogen grid.

Of critical interest:

- This approach avoids the high cost of transporting hydrogen from central facilities to points of use, a problem that has substantially constrained the rollout of hydrogen filling stations and fuel cell electric vehicles in California, and
- When renewable natural gas is used as the feedstock, the hydrogen product will be scored as low-carbon or carbon-negative using GREET-type assessments.

This presentation will describe a demonstration project that is currently underway in southern California, showing how compact process-intensive microchannel reactors and heat exchangers enable the deployment of mass-producible, low-cost "chemical transformers" as hydrogen generators. Deployed in this manner, the chemical transformers enable low-carbon and even carbon-negative hydrogen at affordable costs for transportation and other uses.

Process-intensive hardware enables high-efficient, modular chemical transformers with small footprints and volumes that can be mass-produced at low cost using assembly lines. We project that the levelized production cost of the hydrogen at the point of use, using these systems, will be less than $3/kg, which based on low fossil carbon emissions would be eligible for subsidies – such as the Clean Hydrogen Production Tax Credit – that further reduces costs to hydrogen users.

This technology demonstration project strongly supports the ambition and need to rapidly reduce fossil carbon emissions in order to limit, and then halt, the rise in atmospheric greenhouse gas levels. In the Pacific Northwest, transportation emissions are more than half of all anthropogenic emissions. In order to support the goal of reducing emissions by 50% by 2030, and to support the goals of the 2015 Paris Agreement, we must aggressively establish affordable, clean H_2 infrastructure promptly. These outcomes are achievable through the deployment of Compact Hydrogen Generators as "chemical transformers" on the gas grid, making affordable, clean H_2 available throughout North America and providing a template for others throughout the world.

Participants in the demonstration project, along with STARS Technology Corporation, include Southern California Gas Company, SunLine Transit Agency, the US Department of Energy's Pacific Northwest National Laboratory and RAPID Institute, and several other companies.

AUTHOR INDEX